AUTHOR

Dr John Taylor
John Taylor is a part-time senior lecturer at Canterbury Christ Church University Business School, and runs his own consultancy as a futurist and strategist.

Contents

Chapter 6
Summary of Responsibilities

Chapter 7
Conclusion
Appendix A
Outline of a Disaster Plan

Appendix B
Sources of Information, Support and Advice

INTRODUCTION

We have always lived and worked in the shadow of uncertainty. Will our business be successful in developing and producing the products and services that people will wish to buy and use? Can we sustain our markets and customer base in the face of competition? Will new developments in science and technology render what we do obsolete? Can we recruit and retain the right people with the right skills to deliver our products and services on time and at the right quality?

These questions are the stuff of management. Managers spend their days dealing with the uncertainties that arise as a natural consequence of doing business, servicing consumer demands, and sustaining the market. We are usually quite good at dealing with these kinds of problems, and we have a good idea about what to do when things go wrong.

From time to time, however, some organisations are faced with problems and challenges that emerge from incidents and threats that are significantly beyond the experience and expectations of their managers and employees. When this happens there are no immediate solutions, and often the fact that they survive depends upon a series of happy accidents that enable them to pick up the pieces and recover.

Although the kinds of disasters that may befall an organisation can be quite unpredictable, coping with them need not itself be a voyage of exploration into unknown territory. We can prepare for disaster by having contingency plans that, although we hope they may never be put into action, can speed up our recovery, sustain our business activities, and maintain the performance and morale of our people.

That, in a nutshell, is what this handbook is about. It sets out how to define a disaster plan, describes the steps required to build and maintain it and how to deploy it when required. It concludes by presenting an outline format in Appendix A that may be used by an organisation as a template for developing a disaster plan.

CHAPTER 1

THE NATURE OF THE THREAT

All human activity is beset by uncertainty: what will happen in the future and how people will react to events are essentially unknown variables. Organisations can nevertheless safely plan for many contingencies, in part because of their experience of past events, and also because many situations follow relatively similar trends in the way they develop and in their outcomes. Much of the risk facing organisations in their day-to-day activities can be assessed to some extent, and contingency plans devised to deal with the most likely situations.

Contingency planning is undertaken daily by managers, if only subconsciously, to deal with the unexpected variations in activity in the workplace. Most contingency plans seek to cope with specific incidents, often having limited side-effects on the organisation as a whole or its environment. Indeed the primary purpose of a contingency plan may be to delimit a situation and reduce its impact outside the immediate focus of the problem.

Not all incidents are simple, however, and even simple situations can, with ill-luck, inadequate planning or mismanagement, become complex and far-reaching in their effects. When a situation gets out of control then it may turn into a disaster, that is to say a complex, multi-faceted situation of potentially unknown duration, involving a sequence of events and incidents that are generally beyond the capability of a single organisation to deal with. Some kinds of disasters may demand that the organisation must react under the microscope of public interest and opinion. In other disaster situations the organisation must suffer alone, perhaps because it is very small or because it is a minor player in a much bigger event. When a terrorist bomb explodes it is easy to let pass unnoticed the plight of small shopkeepers suddenly cut off from their customers or their shops by piles of debris and security cordons. The loss of business and the waste of perishable stock is nevertheless a disaster for the small business affected.

Organisations are faced with two kinds of threat: predictable and unpredictable. The relatively predictable threat is a consequence of location, business process, the nature of the product or service and perhaps the nature of the customers and users of the products and services supplied. Some examples of this kind of threat are listed in Box 1.

Box 1: Examples of Relatively Predictable Threats

Regular flooding
Fire
Power failure
Plant and equipment failure
Theft of key equipment
Foreseen disease epidemic
War and unrest in unstable regions

The risks posed by these threats can be assessed in advance and a set of responses may be worked out. In the case of fire, for example, adequate precautions can be identified to reduce the risk and emergency procedures put in place to ensure evacuation of premises and the safeguarding of staff (and sometimes of vital equipment). Relatively predictable threats can be addressed through contingency planning so that procedures can quickly be activated to cope with the issues and their outcomes.

Having a contingency plan means that the procedures can be publicised and practised on a regular basis. For example, all organisations are required to identify fire routes and exits and many organisations appoint fire marshals, who can take charge during an evacuation, both to guide people quickly to safe areas and to ensure that people are accounted for. Regular drills and well-publicised procedures, coupled with training and day-to-day management review, ensure that everyone can be aware of the procedures and the actions to be taken in an emergency.

Unpredictable threats are more problematic and there is a danger either of not being prepared for them at all or of over-preparing for something that may never happen. Examples of unpredictable threats are presented in Box 2. Obviously this list cannot be exhaustive!

The major feature of these risks is that they are unpredictable, and often may appear to be entirely random incidents.

Box 2: Examples of Unpredictable Threats

Flash flooding, freak weather, earth movements
Terrorist attacks, direct or indirect
Civil emergencies
Transport emergencies
Loss of vital services — power, water, sewage disposal
Unanticipated breakdown of key equipment
Sudden death or incapacity of key staff
Assaults on staff
Contamination by chemicals, ionising radiation, or other substances dangerous to health
Major and rapid disease epidemic
Sabotage (including computer hacking)
War and unrest in previously stable regions

Random incidents that are entirely unpredictable can develop from emergencies into disasters. Of course a fire can be a disaster, but we can both insure against it financially and have procedures in place to assist recovery from it. In contrast, a bomb explosion that kills people and destroys property, equipment, goods and materials can have more far-reaching effects, not least in terms of the fear generated in the workforce.

A disaster may have a profound effect on morale and on employee behaviour. For example, in the case of a disaster affecting the journey to and from work, or the families and friends of employees. In those cases the disaster changes people's priorities. People may start to vary their journeys to work, to demonstrate excessive concern for their family (thus affecting concentration on the work itself) and to be concerned about their physical security. For example, a disease epidemic may require employees to nurse members of their family, thus reducing both their availability for work and their morale.

Even if untouched by the primary disaster of, for example, an explosion, an organisation, its suppliers or customers may still find themselves affected by being inside a security cordon. This may prevent trading for an indefinite period.

Whatever the risk and the seriousness of its implications, the management challenge is to be prepared to cope with it. The danger is that in making preparations organisations can go too far, both in the detail of their preparations and in the actions they take to avoid major problems. This handbook seeks to offer a range of actions that are intended to help organisations and their managers be better prepared to face disasters.

Disaster planning should be seen as a separate exercise to other forms of contingency planning, which may deal with more specific threats that, although significant, can be addressed more simply and directly.

Disaster planning for a complex situation of potentially unknown duration is important for several reasons.

- It enables the organisation to define the steps that can be taken to address incidents in an ordered and controlled way.
- It helps to reduce panic by demonstrating that processes and procedures are in place to cope with incidents as they arise.
- It provides a road map for the organisation to get through the crisis and return as quickly as possible to "business as usual".
- It defines the people, places, processes and equipment that are key to the security and well-being of the organisation, both during crises and in more normal times.

CHAPTER 2

PRODUCING A DISASTER PLAN

This chapter examines the steps required to develop a disaster plan and to ensure that it remains up to date and relevant to the changing needs and circumstances of the organisation.

Preparing a disaster plan is a serious exercise that should not be taken lightly. If addressed systematically it can be a valuable activity that not only meets its direct objective but also helps the organisation to have a clearer picture of its scope, operations, processes and activities. In addition it can be a basis for a thorough understanding of the skills and capabilities possessed by employees, which can sustain the organisation through not just a disaster, but many other situations as well.

Box 3: Steps in Producing a Disaster Plan

1. Scoping the plan.
2. Identifying the data.
3. Writing the plan.
4. Walking the plan.
5. Publication.
6. Communication.
7. Revision.

The steps in the production of a disaster plan are set out in Box 3. Before each step is addressed, however, the most important action is for management to determine the roles and responsibilities of both the planners and those responsible for implementing the plan. Without clear definition of roles and responsibilities, it is likely that the process will fail to be taken as seriously as it should be and thus fall into disuse.

For this reason, the initial section on Scoping the Plan begins with consideration of the management process required to ensure that the plan is sustained at a high level of readiness so that it can be deployed with little advance notice or delay.

The keynote of compiling and implementing a disaster plan is simplicity. The documentation should be easily updatable, for example as a loose-leaf manual. It should be clear to read and translate into practice, so it should be concisely written and printed in larger format typeface.

Storage and distribution of the plan is dealt with below, as a special consideration. Whilst the main draft will ideally be undertaken on a computer, the manager of the plan and the key players defined within it should each have a hard copy that can be stored safely and accessibly.

STEP 1: SCOPING THE DISASTER PLAN

This part of the process should deal with defining the nature and purpose of the plan. It should specify the kinds of situations that the plan seeks to address and also the procedure for invoking the plan. Finally it should specify how the plan is to be maintained.

A major part of this step is defining a formal disaster planning team, whose role is to produce, test and maintain the plan. Chapter 3 discusses the role of the disaster management team whose role it is actually to manage any real incidents in accordance with the plan. In many organisations the same people may carry out both roles.

The planning team

The key decisions about the purpose, use, content and format of the plan should be vested in a small group of persons drawn from the main areas of the business. The group should be led by a senior manager at board level who has the power to elicit the information and co-operation required to prepare and maintain the plan.

There should be clear understanding of the order of responsibility within the team and if one member is absent another should be able to step into their place. It is important that the plan is not compromised by the unavailability of a key contributor. These considerations suggest that a minimum of three people should constitute the planning team, representing the main business areas, for example operations, marketing and sales, finance, HR and IT. The team should also be able to co-opt a specialist onto the team such as a safety officer or facilities manager. The responsibilities of the planning team are summarised in Box 4.

Box 4: Responsibilities of the Planning Team

1. Defining the scope and format of the plan.
2. Identifying responsibilities for supplying information and other material for inclusion in the plan.
3. Establishing timescales.
4. Preparation and publication of the plan.
5. Communications related to disaster preparedness.
6. Testing and maintenance of the plan.
7. Managing revision of the plan.

Once the decision has been taken to produce a disaster plan, it should be progressed as an urgent project, with appropriate project management to ensure that it is produced on a defined timescale. Thus the planning team should consist of persons who can dedicate the time and effort required as a priority.

Security

It may appear inappropriate that security is an issue when the aim of a disaster plan is to ensure that the best-possible response is made in a crisis. To ensure that response, however, the plan needs to contain considerable information about the business, including business processes, key employees and their skills and capabilities, lists of customers and suppliers, inventories, key-holders, bank accounts, the location of computer back-up media and any other material necessary to sustain the business. Clearly this information would be invaluable to a competitor!

The easiest way of ensuring security is to issue copies of the plan to named individuals, who themselves would be key players in the aftermath of a disaster. In an extended organisation, each head of department and their deputy should have a copy, to provide extra cover in case any one player becomes incapacitated in the course of the event. In a small organisation, three or four people should each have a copy of the plan. Ideally, while one copy is always kept on the premises, the other copies should be kept in the homes of the holders or, failing that, in separate premises (such as a bank vault) so that destruction of the site does not prevent the plan from being activated.

The names of plan holders should be made available to the organisation's insurance company and to the emergency services to speed the effectiveness of the response. It must be remembered, however, that the emergency services may have different priorities in a disaster and should not be relied upon as primary sources of information or actions in relation to the plan. Actioning the plan is always the responsibility of the organisation itself.

Defining the Purpose and Aims of the Plan

The first task of the planning team is to decide upon the situations in which the plan should be activated. This is far from easy because it implies that disasters and their aftermath can to some extent be predicted, and clearly that is not the case. It is important to define guidelines as to the kinds of situations in which the plan will be activated to avoid its unnecessary deployment when the response would be disproportionate to the events.

Chapter 1 sought to distinguish disasters from what might be termed "regular emergencies", essentially in terms of their unpredictability and potential extent. For each organisation, it is important to analyse what would constitute a disaster in contrast to an incident that could be overcome relatively quickly and easily. The key factors that the planning team may need to consider include:
– the physical location of the site and buildings of the organisation and how that location may be compromised by flash flooding, landslides or other natural phenomena
– the neighbourhood of the organisation in terms of the use of land and property, for example for the storage of dangerous materials or for the conduct of dangerous activities
– the susceptibility of the organisation to the loss of key employees
– the susceptibility of the organisation to the loss of key data or stock
– the nature of the goods and services supplied and of the customer base, in particular where these might provide a source of threat (for example, the provision of goods and services to a particular customer might expose the organisation to terrorist threat, as might the conduct of operations in volatile environments, such as unfriendly or politically unstable states)
– the security of power and fuel supplies, and other key services

– access to and from the site in the event of an incident, and the
    implications of larger scale disruptions of the transport and logistics
    systems
– the problems that may emerge if the business has to transfer to
    temporary premises.

The team will be able to draw upon a variety of support information to
help them in this task, ranging from the contingency planning team in
the local authority to the national security services. A checklist of key
sources can be found in Appendix B.

STEP 2: IDENTIFYING THE DATA

This aspect of disaster planning is concerned with determining the
nature, location and accessibility of key data, rather than with the
gathering and storage of the data itself, although storage is discussed
briefly below. Key data is defined as that which will enable the
organisation to sustain its business in the aftermath of a disaster.

The disaster plan should contain a simple schedule of where to find
the key data. This in effect defines the data that should be treated as
crucial to sustaining business continuity. This circular definition is
deliberate because it is a virtuous circle, thus providing the plan with a
real role in everyday management planning rather than simply
gathering dust waiting for a disaster to happen.

The data required in this stage of the plan may principally consist of
lists, including:
– key-holders, with contact details
– locations of computer back-up media
– authorised persons holding bank account and similar details
– authorised signatories
– administrators of IT and other systems who can override security
    controls (eg to gain access to password controlled files and processes)
– staff with special skills or experience
– staff with specific responsibilities (those licensed or trained to work in
    specialist situations such as on high voltage equipment or gas
    equipment or even those in administrative roles requiring specialist
    training)
– supplier contacts
– customer contacts.

Much of the above information will be subject to the Data Protection Act. People whose contact details are to be included in the plan should give their consent and be aware that there may be an element of risk in their participation. The risk element refers, for example, to people being asked to work at sites that have become dangerous as a consequence of an incident, or which may be a secondary target in the aftermath of an initial incident such as a terrorist attack. It may also mean that, in extreme cases, an employee may become a potential target, as has happened in some incidents related to animal rights. In all cases the organisation should be prepared to indemnify people who are asked to perform key roles in circumstances where other staff have been released from duty.

The planning team should identify who is responsible for compiling each of the areas of information required and the timescales for the supply of the information. It may be the responsible persons themselves who are best placed to define the detail of the composition.

Storing the Data

While the plan can draw upon information contained in other databases and record systems, it may be appropriate, especially in a large organisation, for the plan to have its own dedicated database. This should be small enough to keep separate from the main IT system, so that IT systems failures can only minimally affect the integrity of the plan. One option is to keep the plan data on hard media, such as CD-ROM or DVD, so that it can be accessed using any convenient laptop. Clearly such media must, as a minimum, be password protected and retained in a secure place. At least one copy should be maintained by a nominated person in a separate secure location.

How the plan data are updated will depend on the organisation's IT architecture, but the planning team should have management oversight of the arrangement made.

STEP 3: WRITING THE PLAN

The plan should be written either by a dedicated small team reporting to the planning team or by the planning team itself.

The main considerations during the writing phase are as follows.
1. The plan should be subjected to a project management approach to ensure that it meets the timescale and the criteria laid down for it.

2. Its elements should be tested as widely as possible by those affected, so that a practical focus can be maintained.

It is recommended that the drafts and final version of the plan are carefully dated and version numbered so that there can be no doubt about which version is current.

An example of a plan is provided in Appendix A.

STEP 4: WALKING THE PLAN

It is important that all measures, procedures and actions are properly tested during the formation of the plan and then subjected to regular testing once the plan has been adopted. The procedures to ensure that this happens should be included in the plan. "Walking" the plan will also enable key areas of training and communication to be identified and designed.

Testing the readiness of the plan involves carrying out dry runs, simulations and inspections, as appropriate, to ensure that the plan will work and that the facilities, services and processes invoked by it are ready and available. One easy way of doing this so as not to provoke undue alarm, and to help establish it in the organisation's "psyche", is to include any appropriate parts of the plan in the regular round of fire and other emergency drills. There may be occasions, however, when specific elements of the plan need special attention and training, and these need to be identified and addressed separately to underline their importance.

A key aspect of walking the plan is conducting an "after action review". This can identify areas of poor communication, where procedures need to be improved, where more training may be necessary or where new responses may be required. After action reviews are increasingly used in project management and in the civil and military services. They are an opportunity to consider in detail the steps taken and issues encountered during an activity, so as to identify problems that can be addressed in refining the plan.

STEP 5: PUBLISHING THE PLAN

The security of information that is vital to the well-being of the organisation remains a paramount concern. Nevertheless, it is important that the plan should be published in order that it can be implemented quickly and effectively when an incident is encountered. It is also important, however, that no panic or undue reduced morale results from the knowledge that a disaster plan has been developed. Nor should publication lead to the provision of public information and data in the public domain that could itself place the integrity of the organisation in danger. How the plan is published therefore requires careful consideration.

An easy way of dealing with the conflicting demands identified above is to publish the plan on the basis of "who needs to know", and to handle the wider issues of promoting the purpose, objectives and broad purpose of the plan through a structured communication programme. The outline of such a programme is presented in Step 6 below.

Who needs to know what the plan contains? The key players who need to have access to the plan are listed in Box 5 with indications of the degree of access required.

Box 5: Staff Requiring Access to the Disaster Plan

Job Function	Degree of Access
Board members and senior managers	Whole plan
Disaster planning team	Whole plan
Departmental heads	Summary plus departmental details
Other managers	Summary plus departmental details
Staff in emergency response roles	Summary plus local emergency details
All other staff	Public summary

In small organisations with few staff, it may be appropriate for the whole management team to have access to the plan, with some nominated to have copies in order to fulfil specific roles in an emergency.

With security in mind, it can be useful to provide numbered copies of the plan against a master list held by the disaster planning team. This can also ensure that when a manager leaves or moves within the organisation, their copy of the plan can be transferred to their successor.

STEP 6: COMMUNICATING THE PLAN

The communication of a disaster plan throughout the organisation has two main aims.

1. To reassure all employees that steps have been taken to ensure their safety and security in an emergency or threat situation, and that their livelihoods will also be protected by measures to sustain the organisation's business and activities.
2. To ensure that the emergency procedures and actions are understood and practised.

To avoid unnecessary concern, it may be best to communicate the plan via normal channels, such as team briefings. The official news media in the organisation — notices, in-house newsletters or an intranet — should also be used to emphasis the "business as usual" keynote of the planning process.

The main elements to be communicated include:

– what to do if a threat is received or a suspicious item observed
– who the key players are and a brief description of their roles, such as fire and emergency marshals, first aid workers, security staff, and other nominated emergency staff such as building managers
– locations (including maps) of emergency exits and assembly points
– any facilities for people to reassure their families about their safety (eg provision of emergency mobile phones)
– the need for a continuing high level of vigilance to help identify threats as quickly as possible.

The communication process must also set out guidelines about appropriate behaviour during and following an emergency incident. For example, people should not hold "witch hunts" or engage in amateur detective work to identify suspects. In particular, it is important that people do not focus on persons with unusual names or who appear "different" from other staff. Communication needs to emphasise that threats and emergencies do not set aside the need to comply with legal requirements such as health and safety regulations, the Human Rights Act and the Data Protection Act. Nor does an emergency generate a *carte blanche* for staff to ignore company policies and rules — this is especially important in situations where the organisation must operate from a temporary location.

Customer-facing employees (in fact anyone dealing with people from outside the organisation) may require special information, for example how to deal with visitors during an emergency and how to reassure customers and suppliers that it is "business as usual" as far as possible in the circumstances.

STEP 7: REVIEWING AND REVISING THE PLAN

The plan should be reviewed annually to ensure that it is up to date, although major changes may need to be accommodated as they happen. To achieve this the planning team should meet regularly. The key objective is to ensure that the plan continues to provide a practical solution to securing and maintaining the organisation.

The revision process should review the nature of potential threats. Thus the revision team should maintain close links throughout the organisation to ensure that changes in practices, the locations and layout of new premises and facilities, and changes to the goods and services provided can be assessed. Changes of suppliers, channels to market and any other changes in the nature of the organisation's activities should also be carefully assessed so that new areas of risk are quickly identified.

Regular testing plays an important role in reviewing and revising the plan, both to ensure that the organisation can implement the plan when required and to identify where improvements or changes are necessary. This has already been addressed under Walking the Plan above, but the importance of this process cannot be overstated. Unfortunately it has been demonstrated in real disasters that planned processes have failed to meet the needs of the situation. Not only processes need to be tested, but also equipment and facilities. For example, can people properly use escape routes in an emergency? Are evacuation routes kept obstacle free? Is key equipment actually provided where it is needed? Do employees, and especially those with identified roles in an emergency, know what they are to do, where and what their emergency equipment is, and who will manage them during the incident? Most of these issues are also relevant to normal compliance with health and safety regulations, so most organisations will have experience in dealing with them.

CHAPTER 3

IMPLEMENTING THE PLAN

This chapter examines in more detail some of the main issues and challenges that must be addressed in a disaster plan, with suggestions about how they might be tackled.

When a major incident or disaster happens, there are three steps that need to be taken.
1. Establishing incident management.
2. Identifying business priorities.
3. Identifying key employees.

INCIDENT MANAGEMENT

In preparing for an emergency, it is essential to identify those who will be responsible for managing the emergency as it happens and invoking all or some elements of the disaster plan. The following guidelines are intended to help determine who these key people should be and how they should be prepared and supported.

Disaster Management Team

This is the top level of disaster management in the organisation and may in fact be the disaster planning team in its operational guise. The disaster management team should be responsible for formally deploying the plan and ensuring that it is properly carried out. These responsibilities include:
– ensuring a clear chain of command
– making arrangements for devolving the declaration of an emergency to incident response teams (see below)

17

- managing the establishment, equipping and training of incident response teams — an ongoing role outside emergency situations
- ensuring that incident response teams are properly staffed and in position
- dealing at high level with the emergency services on planning and strategy aspects of the emergency
- dealing with staff representatives and trades unions and other organisations directly concerned with the health and welfare of the workforce (and any other persons present in the organisation at the time of the emergency)
- dealing with the press and public relations.

It may be necessary for additional persons to be allocated to the management team during an emergency, to handle the specific details of these roles and increase the capacity of the team to act quickly over a wide area.

Incident Response Team

The incident response team is a team of managers and practitioners that should be defined at working level and be brought together as quickly as possible to deal with the situation as it unfolds. Ideally they will be people who can easily access the site, either because they live close by or because they on a primary transport route (it should not be assumed that transport will in every case be available). Whereas the disaster management team will be a permanent body (that is, with continuity of membership), incident response teams will be brought together at the onset of an incident and their membership may change during the course of the incident to reflect changing patterns of challenge and needs. The decision to call up incident response team members would be vested in the disaster management team, using an incident directory compiled as part of the disaster plan.

The incident response team should consist of senior managers, with a designated leader and deputies who have full powers in any situation where the team leader cannot be present. The smaller the team the better — ideally three people, with the power to co-opt anyone required to fulfil specific tasks or with specialist skills. The team should be formally appointed, although a proposed member should have the opportunity to decline membership without prejudice given the potential personal risks involved.

If possible team members should not be regular business travellers, although deputies may be appointed.

Local Emergency Teams

Depending upon the size and geographical extent of the organisation, it may be necessary to establish local emergency teams or individuals who can implement and manage the actual emergency procedures in a real situation. In many organisations, the simplest approach will be to use existing fire marshals and first aid trained staff, although they must be aware of the nature of the potential risk and, as with members of the incident response team, have the opportunity to decline. Local emergency teams should report directly to the incident response team, and through that to the disaster management team.

Where an organisation has several sites or buildings, it is a good idea to appoint a building/site emergencies manager (with a deputy) who can assume control in an emergency. Their role includes managing emergency instructions and communications, determining evacuation routes and procedures, securing the site (where appropriate), liaising with the emergency services and providing secure communications with the incident response team. In addition, they will lead the local emergency team at their site.

Although at first sight this approach of having layers of teams may appear to be somewhat bureaucratic, the aim is twofold.
1. To provide a clear chain of command through the organisation, for dealing with issues and ensuring that the disaster plan can be followed even if the normal management chain and communications are compromised.
2. To provide a measure of redundancy in the system in case one or more levels in the management of the incident cannot be activated.

BUSINESS PRIORITIES

A disaster plan should clearly identify the priority areas of the organisation's business processes and activities so that effort can be focused on protecting and defending them. This may mean that employees and other resources can be drafted from areas of lesser importance to protect and sustain vital activities.

The exact nature of business priorities will vary according to the nature and structure of the organisation, and also with current issues. Examples may include:
– work in progress, to meet existing orders and maintain cash flow
– other processes related to cash flow, such as ensuring that invoices are produced and credit is managed

– health and safety procedures.

Organisations may need to work closely with their suppliers and their customers to reduce the impact of an incident on the organisation's wider activities.

One approach may be to appoint an emergency "business in progress" manager who can work across organisational boundaries to ensure that business priorities are quickly identified and addressed. This may be a role to include in the disaster management team.

KEY PEOPLE IN THE ORGANISATION

Alongside the people and teams identified to manage an emergency, it is important to record details of those staff having specialisms or key roles in the running of the organisation. This data will have been collected in Planning Step 2 (see page 11), although it needs to be kept up to date.

The following are examples of key activities that may require specific people to be nominated for emergency duties:
– payroll
– software support
– network management (computers and communications)
– transport management
– electrical power supplies
– operations management (ie the management of production and manufacturing processes).

Within these activities, the aim is to define those activities that must be kept going, if only on a skeleton basis, in order to sustain the business. The information required about each nominated person includes:
– name
– home address and other contact details
– key skill/experience areas
– emergency transport arrangements
– next of kin
– any special needs that may affect performance in an emergency (eg a physical disability that may affect access to an incident site, or a family or carer commitment that may reduce availability).

In addition, there may be people in all areas of the organisation who can provide support in an emergency based upon their previous experience and knowledge. Some people may also possess life skills, gained through hobbies, voluntary work and other extra-curricular activities, that can be drawn upon. For example, a person normally employed as a clerk may have a hobby demanding technical skills, such as maintaining a classic car or railway modelling, and thus be able to provide emergency support.

Perhaps the easiest way of tapping this group is through the line management chain, especially where individuals may be sensitive about sharing their personal knowledge and experience. Line managers should be encouraged to explore these issues with their team and assess whether additional training might help to enhance the use of their additional skills to the organisation.

Note: All information identified in this section must be recorded and maintained in accordance with the provisions of the Data Protection Act.

CHAPTER 4

DEALING WITH SPECIFIC THREATS AND INCIDENTS

This chapter examines the responses that may be appropriate in specific situations. It cannot be overemphasised that the advice given here is intended for guidance — the actual procedures and actions taken will depend on the nature and circumstances of the organisation, the context of the threat, the actual events and the directions given by the police, security and emergency services and others responsible for maintaining civil order. Where an organisation has already developed and tested procedures for use in at least some scenarios, these can be incorporated directly into the disaster plan.

Incidents can affect the organisation directly, such as where it is the target of a terrorist attack, or incidentally, when the organisation is caught up in the aftermath of an event that is focused elsewhere. The situations addressed below are those that involve the organisation directly. Incidents in which the organisation's involvement is peripheral may nevertheless have a serious impact on the organisation's activities, such as when access to a site is prohibited. There will also be circumstances, such as when employees are caught up in incidents while travelling to and from work, or when supplies or the delivery of goods and services are interrupted, that will have significant impact and demand a response. These peripheral situations will be addressed in Chapter 5.

The specific incidents addressed in this section are:

– terrorism
– explosions
– power failures

- epidemics
- severe weather, including flooding
- other serious incidents.

This list is not exhaustive and incidents such as fire emergencies are not discussed here because they will already be addressed in detail by organisations meeting their legal responsibilities under, for example, Health and Safety at Work legislation. The main concern of this handbook is the response to those incidents that occur more rarely and are thus less likely to have been the subject of detailed planning.

There are two elements to the response to the kinds of incidents identified above: a general response and an incident-specific response. The general response has been addressed in Chapter 3 and includes calling the emergency services and establishing incident response teams. The additional considerations in specific cases are as follows.

TERRORISM

Terrorist incidents can take several forms, including:
- placing of bombs
- sending of bombs via letter or package
- personal assault, such as shooting
- hostage taking
- vehicle assaults, such as ramming
- burglary and theft
- sabotage
- computer hacking and related activities, such as spamming
- hoaxes.

Warnings and Threats

A terrorist incident may include a warning, delivered in most cases by telephone, but also possibly by e-mail, fax or even a letter. People who receive telephone calls from the public will need to be trained to deal with a terrorist (or any other threat). This will include people in reception, call centres, service centres and any other person whose telephone number is available in the public domain. It should be remembered that a determined terrorist organisation may also gain access to internal directories, through theft or because of carelessness. The Internet may also be an information source and anyone posting organisational information on the Internet needs to be made aware, through training if necessary, of the implications of including material that may be of use to a malicious reader.

Organisations should consider the following three steps in order to deal with terrorist threats or warnings.

1. Provide training to those employees most likely to receive threats (as described above). This includes how to respond, for example by: calmly seeking to hold callers on the line so that the maximum information about them can be gleaned; by methodically checking any code words or numbers given (ideally getting the caller to repeat them at least once for clarity); and by seeking to ascertain the precise locations of any threatened device or planned assault.

2. Ensure that any equipment used to record calls to call centres (and similar numbers, such as help lines) is operational and in use, so that no threat can be missed.

3. Have in place procedures for the recipient of a call to involve another member of staff (supervisor or manager) who can monitor the situation and alert the emergency services while the threat is being made.

A disaster plan, and communications derived from it, should explicitly emphasise that threats received by telecommunications media, including e-mail, should be immediately notified to the most senior person present. The police should also be immediately informed by dialling 999. If the caller is on the telephone it may be possible to obtain more information about the nature, timing and location of the threat, and any other information about the caller. Even the most apparently innocuous or trivial details may be invaluable in foiling the threat and catching the perpetrators.

Threats made by e-mail should be immediately printed in hard copy, and the e-mail saved. But in no circumstances should any attachment, enclosed or forwarded message or document be opened, for fear that it may include malicious code (eg a virus).

Threats received by other routes not involving direct contact with the perpetrator, such as threats received by post, also require a clear procedure to be in place. Again training is a key factor in this, best cascaded through the organisation via normal management communication to ensure that all staff are aware of the main steps to take.

Evasive Action Following a Threat or Warning

There needs to be a clear procedure for alerting the population at risk of the need to take immediate evasive action.

The simplest form of evasive action is to activate the fire alarm system. In addition, for bomb threats or any other threats that may involve specific locations, as much information as possible about the locations that has been gathered from the caller needs to be passed to incident control. The fastest way of doing that may be via the normal health and safety channels used in a fire emergency.

In the case of a bomb threat, however, a difficulty is that the evacuation may itself bring people into closer contact with the threat. If the threat contains an indication of the location of a device then alternative routes can be devised, but that in turn requires on-the-spot close organisation of fire and other marshals to ensure that people follow revised routes to assembly points. There may be situations where people may be safer gathered in the centre of a building than in, for example, a car park. This is the case where vehicle-based bombs are involved or where a potential bomb location is a car park. The guidance of the emergency services will be crucial in such cases.

Preparation for bomb and similar threats can benefit from regular and in-depth consideration of a range of scenarios, for example to cope with a device placed in a car park adjacent to a main building. Scenario planning, led by the health and safety manager, can help to determine relative risks, identify alternative escape routes, demonstrate the best places for siting countermeasures such as surveillance cameras and identify appropriate access routes for the emergency services. Small organisations that do not have a safety management team may need to use specialist consultants to undertake this work.

Where alternative evacuation routes are identified to cater for different situations, all employees need to be given clear guidance about the specific routes to be followed and the opportunity to be drilled on a range of scenarios. Clear communications systems able to reach all persons on a site are also crucial, and one option suitable for many organisations may be the installation of a public address system that can be easily accessed by the safety or site manager at the onset of an emergency.

Finally, the nature of a terrorist threat also needs to be considered. If the threat or warning indicates the nature or components of a device, then this information needs to be immediately transmitted to the emergency services, in case specialist equipment is required, for example chemical or biological threats. In such rare cases, the best course may be to suspend a normal evacuation if that might result in people becoming contaminated with life-threatening materials. The emergency services will have the prime role in reaching that decision.

Bombs

Bombs may be placed without warning or may be delivered to the organisation via the mail.

The key to dealing with a suspicious package is training, raising people's awareness of the need to be vigilant and to take evasive action should an object be spotted. People often believe that even during times of high alert they will not encounter a problem, and do not wish to make a fuss for fear of being shown to be foolish. This attitude needs countering — people must be encouraged to see that raising an alarm is a matter of being safe rather than sorry. A simple procedure, such as dialling a well-publicised security number dedicated to the purpose, will help people to overcome their natural reserve. Also the organisation itself, through its management, must not be more concerned about receiving hoax calls than dealing with real ones.

In a large organisation, mail rooms and delivery areas are the first line of defence against devices sent through the post or via courier. An organisation that is highly at risk may install sophisticated equipment, such as x-ray screening, but that is beyond most organisations. Thus the training of staff who receive mail and deliveries to identify potential risks is crucial and should include ways of determining a suspect package either from its shape or apparent contents (such as having a strange smell or having wires protruding), or from its ultimate delivery address. Strangely addressed packages may be suspicious, as may packages addressed to individuals who are known to be at risk. For example, a package from an unexpected origin may be suspicious. In all cases, there needs to be a rehearsed procedure for diverting suspect mail to a safe opening facility, the design and operation of which may depend on local circumstances. The emergency services will provide appropriate advice about how to set up and use a safe area and consultants may also specialise in this area.

If a suspicious package is simply found, for example in a doorway or up against a building, it should not be touched, 999 should be dialled, and the area evacuated.

Another way of identifying suspicious mail is to maintain a list of key recipients at risk, as part of the key individuals section of the disaster plan. Such people may include employees who travel on business in volatile regions, or who deal with customers or suppliers who themselves may be at risk (reflecting the business activities of the organisation).

Personal Assault

The main way of avoiding direct physical assault is to maintain a secure access system that requires visitors to identify themselves and the purpose of their visit. Security passes can help in this process but it is important to note the ease with which these can be forged. If this approach is adopted, professional security services should be hired to develop the system. Key-coded (electronic) access to areas of the company can also be valuable and there are many professional suppliers of this equipment. Subdividing the organisation's physical site into key-coded sections can help to reduce threat but may make responses difficult for the emergency services, so their advice should also be sought.

Access controls place security staff in the first line of danger and this means careful security staff selection and training. It is no longer sufficient to retain long service employees in this role simply as a means of rewarding loyal service. Security staff should be trained in the methods of response to be deployed in a situation, the limitations of response and other considerations such as public liability.

If shooting is threatened or occurs, employees should lie on the floor away from windows, ideally under desks or other furniture that may provide some protection. In no circumstances should employees try to disarm or even approach potential armed terrorists — that should be left to the security services.

Hostage Taking

People who are most at risk of being taken hostage are senior managers who have access to financial systems (including money), security systems (including building and computer access codes) or materials of direct use in terrorism, such as explosives or timing devices. The disaster plan should identify those most at risk and there should be contingency plans to remove them to a safe place at the onset of an incident. Other people may still be at risk from random seizure as hostages, if only to be used as bargaining counters.

The key response is to summon the security forces (via the police) as quickly as possible. The security forces are well versed in dealing with hostage incidents, and the organisation's main concern will be to sustain business continuity, which is addressed in Chapter 5.

Vehicle-based Assaults

The use of vehicles to ram buildings in order to breach security (eg to support theft, to attack individuals or to deliver a bomb) can be countered by having good access controls to the site to ensure that vehicles cannot approach buildings at close quarters. This can be achieved by placing concrete or other barriers that physically bar access to vehicles.

A suicide bomber poses the ultimate threat because they will have no concern about their own life or about causing physical or collateral damage, so access control points need a "hotline" or "panic button" to activate a security response. The procedure to be followed should be included in the disaster plan.

Burglary and Theft

Terrorists may commit burglary or theft to gain access to money or physical resources. There are few additional actions that can be taken to deter such activities beyond those that many organisations will already have in place, such as surveillance systems and more rigorous access control and secure storage. Valuable items such as computers that can be sold for cash should also be specially protected, and it is appropriate for organisations to consult their insurers as well as the police for advice in specific cases.

Another action that is increasingly an option for monitoring and tracking the movement of goods and materials is to mark them in ways that are not easily confounded. This is an extension of the existing technology of etching vehicle window glass with the vehicle registration number, the idea being to deter criminals by making it harder for them to trade stolen goods. Whether terrorists would be so easily deterred is questionable, but sophisticated electronic tags that can increasingly deploy global positioning capability may work well for high-value items such as computers or sensitive materials such as explosives, detonators and timing devices.

Sabotage

Sabotage is a complex area because it can have several underlying causes. It may be a purely terrorist act, but it can also be an act committed by disgruntled employees, working either alone or in concert, for a variety of reasons.

In the case of an externally provoked attack that can be construed as terrorist in origin, there is also the issue of whether the attack is carried out by people from outside the organisation who gain access to vital equipment or facilities or whether it is an "inside job".

Protection against an invasive threat follows the broad approach already described relating to burglary, involving the use of surveillance and other procedures. Protection against an "inside job" is less easy because it may not be possible to predict who may be susceptible to terrorist propaganda, and even then they may not be motivated to translate their beliefs into actions. Line managers can have a key role here, by paying attention to the views and opinions of their staff and noting any patterns that may be suspicious. This path is dangerous, however, because it may infringe the human rights of employees; managers should not seek to keep dossiers on their staff, especially if they restrict their attentions to staff with apparently foreign names or origins. The best course of action may be to share their concerns with a senior manager so that there can be a level of awareness that may be followed up if a situation develops. It may be appropriate to nominate a senior manager in the disaster plan who can receive the suspicions of managers down the line and who can act should apparently worrying patterns emerge. The information should be passed to the police to investigate fully.

It may be appropriate to discuss the intelligence-gathering process as a concept with trades unions and other employee representatives, so that there is no question of it being seen as a way of conducting witch hunts or to victimise individuals.

Sabotage undertaken by disgruntled employees may be addressed along similar lines, with any emerging suspicions about intended or planned attacks being passed to the police.

Computer Hacking

Malicious attacks on computers can take several forms, including direct attempts to destroy equipment or software, and misuse of software to gain access to sensitive systems or records. In turn this can lead to systems or facilities being effectively sabotaged either to stop them working, or to make them work in new ways that benefit the aims of the attacker. Alternatively, computer misuse can be undertaken to gain unauthorised access to records and data that may be then used against organisations or individuals, perhaps to divert funds, or to gain intelligence to damage the organisation's interests, or to identify individuals who may then be placed at personal risk. The proper use of firewalls and other protective software, and careful management of passwords and other security processes can guard against many attacks. Equally important is the control of computer-generated data, such as printouts. Clear guidance about the management of passwords and the storage and disposal of data should be communicated throughout the organisation, supported by training, and underpinned by regular audit procedures that can ensure that the procedures are being carried out properly. Detail of the practices and procedures outlined above should be included as appendices to the disaster plan.

Hoaxes

It should not be assumed that any warning or threat received, or suspect item detected, is a hoax. The activation of emergency procedures should in all cases take place as set out in the disaster plan, as communicated within the organisation at large.

If, subsequently, an incident is discovered to have been a hoax, then the emergency/security services will take appropriate steps for those responsible to be apprehended for what is a criminal act. If it turns out that the perpetrators are employees of the organisation, then the normal disciplinary procedures should be applied.

EXPLOSIONS

There are many sources of explosions that can be a disaster for an organisation, and the main issue for disaster planning is dealing with the aftermath. This is addressed in Chapter 5.

Knowledge of the risk of explosions can help organisations to make specific plans, however, and can be gained from an assessment of the physical environment of the business. For example, the nature of surrounding industry, the presence in the locality of risk-generating installations (such as oil storage tanks), and the undertaking of hazardous activities (such as the use of a nearby railway or road for the regular transport of dangerous goods), can all lead to a higher level of risk. The emergency services will be able to give advice in particular circumstances of any planning or preparatory measures that an organisation should take to reduce the risks that it faces, and to deal with the aftermath of an incident.

POWER AND FUEL SUPPLY FAILURES

A sudden and sustained failure of power or fuel supplies that lasts longer than the organisation can easily maintain its normal activities constitutes a disaster for many businesses. The disaster plan should include estimates of the supplies needed to maintain a skeleton operation and the ways in which supplies may be sourced at very short notice, such as through the use of emergency generators. While computer data can be protected provided that there is a proper back-up regime in place that is also properly used, heavy machinery and equipment may have to be suspended and that can lead to problems meeting the demand for goods and services. The key issue then becomes maintaining business continuity, which is addressed in Chapter 5.

Disruption of vehicle fuel supplies is a special case, and the disaster plan should specify which vehicles are to be treated as essential. In addition, any special arrangements for transporting essential employees to and from work should be specified, such as the hiring of buses or other vehicles.

EPIDEMICS AND PANDEMICS

Serious outbreaks of life-threatening diseases should become known through the media and other news sources, and government decisions about protection and treatment regimes and the movement of people will be publicised. The disaster plan will identify key people who may be put forward for protective treatment (eg vaccinations) if treatment supplies run low. All other employees should be advised to seek vaccination (or other recommended treatments) and, in line with any official advice, be counselled to stay at home. At the same time external visitors to the organisation may be discouraged. In a retail establishment the position is more complex, and official advice may be sought from the local medical officer or the organisation's own medical consultant.

WEATHER EMERGENCIES

Weather emergencies such as storms, tornados or floods should be forewarned in weather forecasts, but history suggests that details such as timing and extent may not always be reliable. As part of disaster planning, some simple steps can help to stave off serious problems or at least reveal areas at risk.

A site that is well-stocked with trees should be inspected by a tree surgeon at least annually to ensure that storm damage from flying branches, or whole trees becoming uprooted, can be minimised.

A review of the site location and layout can reveal areas at risk from flooding so that emergency measures can be taken, such as having a stock of sandbags.

Similarly, structural surveys of buildings can reveal areas of risk that may become serious in a weather emergency (and these should be treated when discovered, not when a storm is predicted!).

A key contribution to planning in this area can be made by the local planning authority at either district, city or county levels. These bodies have a statutory duty to prepare contingency plans for civil emergencies and can therefore give advice to organisations having premises or other interests in their areas.

OTHER CIVIL EMERGENCIES

The range of potential civil emergencies is large, ranging from major traffic incidents to plane crashes and other unpredictable events. In most cases the emergency services and the local authorities will provide direct advice and support, and also give direction about specific actions to be taken. The value of undertaking "What if?" scenario planning has already been mentioned, and this process can include inputs from the emergency services as well as from internal management and health and safety specialists. The Health and Safety Executive (HSE) is one of a number of specialist agencies that will provide advice through publications and on site. Contact details for the HSE and a number of other agencies are provided in Appendix B.

To conclude, it must be stressed that the guidance and advice presented is by no means exhaustive, either of the types of incidents that may be encountered or of the responses that may be required. Much will depend upon the products and services provided by an organisation, the scale of its operations and the environment within which it is situated.

CHAPTER 5

MANAGING THE AFTERMATH OF AN INCIDENT

This chapter discusses the actions to be taken to manage the aftermath of an incident. These actions should be specified in a disaster plan so that a co-ordinated approach can be sustained. They fall under two headings:
– emergency procedures
– sustaining business continuity.
The emergency procedures to be addressed in the disaster plan have to an extent already been described in Chapter 4. This chapter therefore focuses on the additional steps required in the period immediately following the event.

EMERGENCY PROCEDURES

The disaster plan should include procedures for the following:
– liaising with the emergency services
– evacuating people at risk
– dealing with casualties
– providing support for victims
– securing the site
– preserving evidence for subsequent investigation.

35

Liaising with the Emergency Services

Although the immediate response to an incident may involve dialling 999, a disaster plan should include telephone numbers and other contact points that should be used subsequently to deal with the emergency services. These arrangements may, of course, be supplanted by the emergency services themselves.

The plan may also provide details of the nature of the liaison with the emergency services, such as providing nominated contact points for specific purposes. The precise arrangements will depend upon the nature of the organisation and its activities. An example would be for contacts on both sides to be nominated with respect to dealing with hazardous substances.

This section of the disaster plan should be compiled with the active support and contribution of the emergency services and other key agencies, such as the local government authority.

Evacuating People at Risk

This issue has already been addressed in Chapter 4. The key points, in summary, are to use existing arrangements for evacuation in a fire emergency, but to be ready to vary them on the advice of the emergency services.

If a threat has been identified that compromises a defined evacuation route or assembly point, then a local decision will be needed to vary the route. The disaster plan should specify the command structure for achieving this (in effect, it will be the responsibility of the nominated fire marshals working under the control, respectively, of the site manager and the local incident response team).

Dealing with Casualties

Following a disaster, there is a high likelihood that there will be casualties on a scale beyond the normal experience of the organisation's own first aid teams. Nevertheless, the disaster plan should contain a list of trained first aid volunteers who may in some cases need to be brought to the site. The plan should make clear, however, that first aid provision should come under the direction of the emergency services once they arrive on site. The emergency services may have to undertake on-site triage (categorisation of injuries) and a vital role of first aid staff may be to comfort patients during that process.

It is likely that there will be volunteers (who may be members of the public as well as employees) who come forward to help those in shock or with minor injuries, and the plan should specify how volunteers should be managed. Ideally, an emergency place of safety should be identified where victims may receive comfort and refreshments while they are being treated or are awaiting transfer from the site.

The disaster plan should also indicate how emergency transport of victims from the site can be arranged, not only using the organisation's own vehicles but using volunteers and commercial suppliers, eg by providing the contact details for taxi services.

Providing Counselling and Support for Victims and Others

In the aftermath of a disaster or a major incident, counselling services are increasingly being made available by the authorities. The organisation may itself wish to support employees and their families who have been casualties or victims, and the plan should set out how this is to be achieved so that it can be quickly activated. The kinds of support that may be provided include visiting injured employees and their families, supporting bereaved families and making arrangements for return to work and ongoing support in the workplace.

A specific requirement may be to provide support for employees who lose colleagues through death or serious injury, whether in an incident at work or in a disaster elsewhere, such as during a journey to work. Sometimes the first indication that someone has become a victim is that they do not arrive at work, and that itself may be a source of shock and stress for their colleagues. The counselling required is very similar to bereavement counselling.

One approach that can be pre-planned, but is perhaps only suitable for a large or high-risk organisation, may be to have a standby arrangement with the clinical psychology or psychiatric services provided by the local NHS authority or a private health supplier. Such arrangements should be planned in close consultation with the supplier.

Securing the Site

The disaster plan should make it clear that the primary issue in a disaster is the protection of life, with the security of property being of secondary concern. Nevertheless, once the site has been evacuated then steps should be taken to protect it, for three reasons:

- to ensure that exposure of third parties to dangerous materials or wreckage is reduced to a minimum
- to protect evidence that may be of value in any investigation from interference
- to secure the property from theft or vandalism.

The disaster plan should specify how site security is to be achieved, under the overall authority of the disaster management team and in accordance with the advice of the emergency services. This may include providing contact details of any commercial security organisations that are to be deployed.

It may also be necessary to deploy employees to site security duties. Ideally this should be a volunteer process, although the disaster plan may include contact details of known individuals who are willing to act in an emergency.

It must be made clear in the disaster plan that all site security activities must be operated in accordance with health and safety requirements, and that those involved also have joint responsibility for the safety of themselves and others. The plan should ensure that there is a continuing health and safety presence at the site to protect both individuals and the organisation from hazards.

Preserving Evidence for Subsequent Investigation

The emergency services will determine the actions to be undertaken to investigate an incident, and they will involve the security forces as appropriate. The arrangements that the organisation may have to make to support this may include:

- securing the site (see above)
- managing access to the site
- providing records and other material related to the onset of the incident (such as records of telephone calls, e-mails and statements from persons involved).

Where possible the identities of people who have played a major part in an incident, whether as participants in the process or as key witnesses, should be preserved and made available to the investigators as required.

In the case of sabotage or attempted sabotage that may threaten life, the first concern is to evacuate the area, following the same procedure as with a bomb threat. The area around the site or incident of sabotage will need to be secured in order that the police and other services may conduct forensic examination after the event.

Lists of persons having access to the area prior to the sabotage should be secured, along with any physical records such as video recordings, to help with subsequent enquiries.

As soon as the initial response, including any evacuation, has been implemented, managers and specialists (eg technicians) with direct knowledge of the equipment, facilities or services affected should be assembled to help both with the immediate response and subsequent enquiries.

SUSTAINING BUSINESS CONTINUITY

The degree to which the activities of the organisation can be continued in the aftermath of an incident will depend on the nature and extent of any damage sustained or the loss of key materials such as supplies and finished goods. The immediate decision as to whether normal activities can be resumed will be taken in accordance with the directions of the emergency services, taking account both of safety and of the need to preserve evidence for an investigation. The decision to restore activities should be taken by the disaster management team.

Faced with significant barriers to conducting its business on a day-to-day basis, an organisation needs to have a range of fall-back options that will enable it to sustain its operations whilst recovering its business position. The actions that need to be taken may include:
- finding and setting up alternative premises
- arranging for staff to be transported to temporary premises
- ensuring that customers and suppliers are aware of the temporary arrangements
- arranging for customers to reach the temporary site, including providing transport facilities if necessary
- ensuring the availability of communications — mail, e-mail and telephone
- monitoring the recovery of the main site so that the period of use of temporary premises can be kept to a minimum (in order to reduce costs).

Core Processes

The minimum activities required to be undertaken to sustain the supply of goods and services should be identified in the disaster plan.

Depending upon the status of inventories at the time of the disaster, it may be possible to continue business by running down stocks. To do this requires an assessment of stock levels and work in hand. In turn this will determine the minimum production requirements necessary to meet demand.

Staffing

The major requirement will be to define the numbers, skills and duties of the people who will work at the temporary site. In general it will not be possible to name all the staff in the plan, but key managers, specialists and skilled workers should be identified where possible so that they can be transferred to a new site with minimum disruption.

Temporary Premises

The work level, storage of stock and numbers of staff identified will determine the space requirements of a temporary site. The key issue is whether any manufacturing or assembly work will need to be done.

Office space is relatively easy to obtain through agencies and specialist office rental companies. The plan should include the contact details of such agencies, with indications of locations, lead times of availability and indicative costs.

One possible option may be temporary help from neighbouring companies, who may be able to provide short-term accommodation for office work and sales. Suppliers and business customers may also be anxious to help, if only to help preserve their own positions.

While office space can be used for a range of work, for example desk-top assembly, design, packing or small-scale storage, it is not generally suitable for manufacturing on anything but a small scale. The plan needs to identify what manufacturing (including major assembly work) is required to sustain the business, and hence specify the following:

- space requirements
- floor loadings
- power demands
- access demands
- health and safety considerations
- transport of staff between existing and temporary locations
- transport arrangements for the delivery and collection of goods and materials

– communication links.

It is highly likely that owners of temporary premises will not wish structural alterations to be made, and in any case these will take time and financial resources and should be avoided if possible.

CHAPTER 6

SUMMARY OF RESPONSIBILITIES

This chapter is intended as a reminder of the range of responsibilities and requirements that must be undertaken both during planning and in the management of an actual incident. Some of these are statutory requirements, while others are good management practices. Observing them will help to avoid problems emerging once normality has returned. Please note that because legislation and practices are always liable to change, the following advice is for guidance only, and does not seek to be a definitive representation of the law.

The key guidance is that unless in an exceptional circumstance civil legislation is suspended by an Act of State (which can happen in some kinds of emergencies), it must be presumed that all legislation continues to have its normal and full effect. Thus all emergency procedures should be tested in terms of their compliance with the law.

Specific areas of legislation that may affect an emergency response include the following.

DATA PROTECTION

All lists and details of key personnel are subject to data protection legislation. Personal details must be stored confidentially, must be kept up to date and must be retained only for so long as is necessary. It is also necessary to put in place a procedure to ensure that legitimate requests for information about the data stored are complied with in accordance with statutory requirements.

HUMAN RIGHTS

A frequent feature of emergency responses is the need to deploy people in areas that are different from their normal roles, perhaps at very short notice and in ways that may disrupt their family and social arrangements. Nevertheless that does not mean that a person's human rights can be compromised, although clearly there may be many people who will voluntarily set aside their normal concerns to undertake urgent actions. It is important to ensure that a person is not subsequently disadvantaged by the roles or actions they are asked to perform in an emergency. Careful planning can help to avoid problems in this area.

Regardless of the need to comply with data protection legislation, it is good practice to be able to demonstrate that individuals who are identified for particular roles in the aftermath of an emergency are those most suited to the roles concerned. This will usually mean that they will have specific skills or experience that can be called upon in an emergency. In this way, although it may be necessary to suspend normal procedures for selection, appointments and promotions during an emergency, it can be argued that the summary actions taken to protect the organisation have as far as possible complied with legal requirements to avoid discrimination in the workplace. Steps should be taken after the emergency to review and ratify any selection and appointment decisions made, so positions can be regularised where appropriate. In particular it may be appropriate to deem all appointments to higher positions made during an emergency to be temporary in nature, and subject to later review.

Where people are clearly identified in plans and procedural documents because they have skills or experience of value in an emergency, it is important to ensure that they are fully aware of any dangers that may result, and have the opportunity to decline a role. If direction of labour becomes necessary, for example to protect vital equipment and services or to meet the demands of the security services (who may have assumed emergency powers), then there should be a formal procedure that can be followed for the identification and direction of individuals. This procedure should form part of the planning process, to support the identification of key processes and equipment "at risk".

A special case emerges where people are scheduled for maternity or paternity leave when an incident occurs. In general, key jobs should already have been covered, and those arrangements may continue. Alternative arrangements may be necessary to deal with urgent redeployment of resources, but in no circumstances should someone on maternity or paternity leave be required to return to duty. On the other hand, a person on paternity leave may volunteer for work, and each case can be treated on its merits (but no-one should be pressured to volunteer).

In no circumstances should management seek to take certain unilateral actions that may breach human rights and related legislation even if to do so may seem legitimate to protect life and property. Specific actions that must not be undertaken include:

- keeping records of employees, or others, who are considered "suspect" on the basis of criteria such as names, ethnicity, or perceived religious or political stances or beliefs
- undertaking surveillance of individuals except in situations where the aim is to protect site security or to guard against illegal activities such as theft
- discriminating against individuals on the basis of suspicions aroused by their names, ethnicity, or religious or political stances or beliefs.

Those who overhear conversations, or who in the course of the normal monitoring of communications (such as e-mails and telephone calls) made using the organisation's facilities and equipment identify suspicious behaviour, should report their observations to management. In turn, management should have arrangements in place to notify the police about issues of concern. All records of such incidents should be handed to the police if requested, and no other actions taken except under the direction of the police. Once any investigation is completed, any retained information should be destroyed (unless otherwise directed by the police).

HEALTH AND SAFETY

Unless otherwise directed by the security services, or as a consequence of emergency regulations issued by the authorities, it must be presumed that health and safety legislation will remain in full force. In a situation such as an explosion, however, where damaged buildings may trap individuals, the saving of life must take priority over the strict application of rules and regulations. Nevertheless, if all steps are taken in the circumstances to proceed in a safe manner, there should be no problems. In an isolated incident it may be possible quickly to involve representatives of the Health and Safety Executive, but that may not be the case in a wider emergency.

In the immediate aftermath of an incident there may be a range of volunteers, including members of the general public, who come to help, especially where saving of life is involved. While help should not be turned away, the site owner or tenant and their representatives will continue to be responsible for the health and safety of everyone on site. All steps must therefore be taken to ensure that volunteers are managed in ways that reduce the risks to which they are exposed. This issue may need to be specifically addressed in the disaster plan, with one or more managers identified having clear responsibility for the organisation of volunteers.

CIVIL EMERGENCY IMPACT

Local authorities have a statutory responsibility for civil contingency planning, drawing together the roles and activities of a range of emergency and security services. They may give specific direction about the actions an organisation must undertake.

NATIONAL SECURITY

In situations where an incident may threaten aspects of national security, the security services will give direction as to the actions and activities to be undertaken.

CHAPTER 7

CONCLUSION

This handbook has sought to describe the planning activities that will help organisations to cope with disasters and similar incidents that are generally beyond the internal resources of the organisation to manage and control.

It is a complex field, with a wide range of incidents and contingencies that may need to be identified and addressed. Also, the ways in which an organisation may be affected by an incident will depend upon the organisation's structure, geographical location, deployment of human and other resources, and the goods and services that it provides. Nevertheless the handbook presumes that, however complex the organisation, there are still significant steps that can be planned which will mitigate the impact of the incident.

The handbook is likely to be of particular use to many small and medium-sized organisations that hitherto may not have seen formal disaster and contingency planning as a priority. It is hoped that the issues addressed will raise the profile of disaster planning as a significant concern for organisations of all kinds.

Finally, any handbook in this field faces the problem of keeping up to date with changing events and approaches to coping with them. Users are therefore urged to inform the author, via the publisher, of any key omissions, or of areas requiring review, so that future editions are not only up to date but also address the changing key concerns of organisations facing potential disaster situations.

APPENDIX A

OUTLINE OF A DISASTER PLAN

At the end of this outline several annexes are provided as illustrative specimens and templates. Their precise form will vary for each organisation and its circumstances. References to some annexes are indicated as ".........." and are not supplied as examples because they will be specific to each organisation and incident. A separate page for each annex is advisable for clarity.

ABC COMPANY: DISASTER PLAN

Date/Version Number [   ]

Scope

This plan relates to the business and operations of ABC Company. It details the procedures to be followed in an emergency situation and the actions to be enacted to protect and sustain the employees, the business, and the customers/suppliers of the company.

Management of the Plan

The plan has been prepared and is sustained by the Disaster Emergency Planning Committee, consisting of the following:

- Managing Director
- Works Manager
- HR Director
- Marketing and Sales Director
- Disaster Plan Manager
- Secretary

Details of the current incumbents of these positions, along with their nominated deputies, are in Annex A. In all cases the deputies have full powers when exercising their deputising roles.

The plan was developed in [month, year] and will be reviewed at least annually to take account both of changing circumstances and of changes in the nature of perceived threat.

The Disaster Plan Manager has the prime responsibility for sustaining and developing the plan, and for undertaking any day-to-day activities required to sustain the readiness of the business to cope with an emergency.

Disaster Organisation

Every part of the company is required to have a local response team capable of dealing with threats and incidents as they arise in accordance with this plan. In addition, each team must be prepared to use its own initiative on those occasions where communication with the disaster management team is not possible or where the speed of a response is paramount to protect people and property.

The current Disaster Organisation Structure is set out in Annex B.

Incident Response

Primary Response to a Threat or Suspicious Action

If a threat to life or property is received from any quarter by any means, a suspicious object is detected, or persons are suspected of acting in a manner that may lead to an attack on persons or property, the procedure to be followed is stated at Annex C. This procedure is mandatory.

Responsible Persons and Emergency Contacts

The main persons having responsibility in an emergency are set out in Annex D. This information includes details of names, organisational addresses and contact numbers and home addresses and contact numbers, and thus must be treated as confidential except during an emergency under the direction of the disaster management team.

Contact information for the emergency and other support services are contained in Annex E.

Key Processes

The main activities and processes are set out in Annex F. Each has been assessed as crucial, valuable or useful, to indicate their relative importance in the event of a disaster threatening the continuity of the organisation's activities.

Crucial activities must be sustained as normally as possible in the aftermath of a disaster.

Valuable activities should be sustained where possible, but can be limited and undertaken on longer timescales than normal. Line managers will review these activities in consultation with local response teams.

Useful activities should be deferred where possible, ideally until after normal business has been resumed. Again, line managers will review these activities in consultation with local response teams.

Special Areas of Risk
Site B has been identified as being on a flood plain and the steps to be followed on receiving a threat of flooding are set out in Annex .... The Local Authority Civil Contingencies Unit will assist once an emergency is established.

Aftermath Management

Specific arrangements for dealing with victim support, employee assistance and employee counselling are contained in Annex ....

[Additional sections dependent upon the nature and circumstances of the organisation.]

Sustaining Business Continuity

Arrangements have been made with VWX organisation for the supply of temporary office space, furnishings and fittings, and emergency equipment such as PCs. [Annex ....: local response team supplies details].

[Additional sub-sections dealing with transport arrangements, key employees, emergency telecommunications services and so on.]

ANNEX A: DISASTER ORGANISATION STRUCTURE

Disaster Emergency Planning Committee

Managing Director: [Name, Contact Numbers, Deputy, Contact Numbers]
Works Manager: [insert details as above]
HR Director: [insert details as above]
Marketing and Sales Director: [insert details as above]
Disaster Plan Manager: [insert details as above]
Secretary: [insert details as above]

ANNEX B: DISASTER ORGANISATION STRUCTURE

Disaster Emergency Planning Committee

[Organisation diagram, names and contact details]

This is a permanent team chaired by the Disaster Plan Manager and representing all functions. Specialist members include the heads of Health and Safety, Site Security and Estates Facilities Management.

Disaster Management Team

[Organisation diagram, names and contact details]

This may be the Disaster Planning Team in its operational format.

Incident Response team

[Organisation diagram, names and contact details]

This team will be chaired by a senior manager with representatives of all functions. Specialisms represented will include Health and Safety, Site Management and Site Security. It may include employee representatives.

Local Emergency Teams

The composition of each Local Emergency Team will depend upon the nature of the threat or incident, the availability of people and the need for specialist advice. In all cases those persons involved should know their roles and responsibilities in advance (as part of the planning process) and have received any appropriate training. An example of team composition may be as follows.

Building/Site Manager (leader)
Site Safety Manager
Works/Operations Manager
Specialist staff

ANNEX C: PRIMARY RESPONSE TO A THREAT OR SUSPICIOUS ACTION — MANDATORY PROCEDURE

If a threat to life or property is received from any quarter by any means, a suspicious object is detected or persons are suspected of acting in a manner that may lead to an attack on persons or property, the following steps must be taken. This procedure is mandatory.

1. Dial 999, ask for police and give the details requested. The police will determine what other emergency services will be required to attend.
2. Inform the immediate line manager or any manager in their absence.
3. Do not interfere with any package or device that may be suspicious.
4. Be ready to evacuate the premises at short notice. Treat the evacuation as if it is a fire emergency.
5. If there is any fire associated with the incident, operate the nearest fire alarm and leave the area.
6. In all cases of an evacuation, attend your normal assembly point and remain there until told otherwise, in accordance with the prevailing fire emergency procedures of the organisation.

Any variations to the above procedure may only be operated under the direct instruction of either a member of the Local Response Team or a member of the emergency services.

ANNEX D: DISASTER MANAGEMENT PERSONNEL

The following is a specimen table of responsible persons.

Role	Name	Organisation Address	Contact Nos	Home Address	Home Contact Nos
Disaster Manager	Name 1				
Health and Safety	Name 2				
Building X Manager	Name 3				
Building Y Manager	Name 4				
Building Z Manager	Name 5				

This annex should also contain lists of the following.

Key holders
Authorised persons and signatories
Administrators of IT systems, with deputies
Specialist/key employees
Key supplier contacts
Key customer contacts
Location of IT back-up systems

ANNEX E: EMERGENCY CONTACT NUMBERS

Police:	Immediate 999	Follow up:
Fire:	Immediate 999	Follow up:
Ambulance:	Immediate 999	Follow up:
Bomb Disposal:	Immediate 999 (Police)	Follow up:
Coast Guard:	Immediate 999	Follow up:
Local Authority Civil Contingency Unit:		

Specific Incident Emergency Numbers

The following is provided by the local response team.

Water emergency:	
Electricity supply:	
Gas supply:	
Local hospitals:	
Emergency transport:	
Taxi services:	
Tree surgeon:	
Victim support and counselling:	Provided by Local Response Team
Next of kin enquiry numbers:	Provided by Local Response Team
Other public enquiries:	Provided by Local Response Team

[This list is not exhaustive]

ANNEX F: KEY ORGANISATIONAL PROCESSES AND ACTIVITIES (EXAMPLES)

Crucial Activities

These must be sustained as normally as possible in the aftermath of a disaster.

Function	Process/ Activity	Responsible Person and Deputy	Contact details
Finance	Payroll	Mr A Mrs B (Deputy)	
	Customer Billing		

Valuable Activities

These should be sustained where possible, but can be limited and undertaken on longer timescales than normal. Line managers will review these activities in consultation with local response teams.

Function	Process/ Activity	Responsible Person and Deputy	Contact details
Operations	Preventative Maintenance		

Useful Activities

These should be deferred where possible, ideally until after normal business has been resumed.

Function	Process/ Activity	Responsible Person and Deputy	Contact details
HRM	Performance Management Support		

APPENDIX B

SOURCES OF INFORMATION, SUPPORT AND ADVICE

BUSINESS LINK

Website: *www.businesslink.gov.uk*

CABINET OFFICE

Civil Contingencies Secretariat
10 Great George Street
London SW1P 3AE
Tel: 020 7276 5053
Fax: 020 7276 5113
E-mail: ccact@cabinet-office.x.gsigov.uk
Website: *www.ukresilience.info*
Information and advice about the Civil Contingencies Act 2004:
www.ykresilience.info/ccact/

DEPARTMENT OF HEALTH

Information on epidemics and pandemics, including contingency plans,
can be found at *www.dh.gov.uk*

HEALTH AND SAFETY EXECUTIVE

Civil   contingency   website: *www.ukresilience.info/contingencies/*
cont_index.htm

MI5

The Security Service maintains a general site advising about the nature of threats and also provides specific advice to business.

Main site:	*www.mi5.gov.uk/*
Business advice:	*www.mi5.gov.uk/output/Page5.html*
Good practice:	*www.mi5.gov.uk/output/Page167.html*
Bomb protection:	*www.mi5.gov.uk/output/Page37.html*

LONDON CHAMBER OF COMMERCE

PDF file at website: *www.londonprepared.gov.uk/business/ lcc_disaster_recovery.pdf*

NATIONAL PRESERVATION OFFICE

A good example of disaster planning related to libraries and collections, which can be adapted to any kind of organisation, from the British Library: website: *www.bl.uk/services/npo/disaster.html*

EMERGENCY PLANNING SOCIETY

Website: *www.emergplansoc.org.uk/*

INDEPENDENT CONSULTANTS

There is a wide range of consultants that can be located on the Internet using the search keywords "business disaster planning" or similar. The list is too large to reproduce here.

US WEBSITES

Some US sites carry information that may be relevant to UK organisations. An example is Blue Claw: *www.blueclaw-db.com/ disaster_recovery_plan.htm*

INDEX

Would you like more copies?

Subscribers to *Croner's Reference Book for Employers* will receive this publication free, as part of their subscription.

Extra copies are available at the following prices:
- single copies — £20 each
- 2–9 copies — £15 each
- 10+ copies — £10 each

To purchase extra copies of *Disaster Planning* simply:

Telephone: **020 8247 1632**
Email: **sales@croner.co.uk**
Visit **www.croner.co.uk/disaster-planning**

Please quote promotion code UQ/P025220.

Human resourceful

To be resourceful you need access to the right resources. This is particularly true for anyone dealing with HR issues as new legislation is coming into force all the time.

The good news is that there is a powerful online resource that gives you instant access to all the facts with straightforward interpretation you can act upon.

CRONER-i Human Resources
- In-depth online coverage of HR topics
- Telephone advice line
- Step-by-step guides
- Model policies – downloadable and ready for use
- Employee fact sheets for your staff
- Sample forms and letters
- Legislation tracker
- Latest news
- Hot topics

The even better news is that you can enjoy **CRONER-i Human Resources** on a five-day FREE trial – with no obligation to buy.

Register for your free trial now at
***www.croner*.co.uk/hrtrial**

Danger Zone

Keeping your staff and business safe means having access to information you can rely on. This is particularly true for anyone dealing with health and safety issues as new legislation is coming into force all the time.

The good news is that Croner has launched a powerful online service that gives you instant access to all the facts with straightforward interpretation you can act upon.

CRONER-i Health & Safety Expert
- 30 key British standards
- EU directives
- Over 100 health & safety topics
- 40 full text Acts
- Legislation tracker
- Personalised weekly e-alerts
- 50 of the most commonly used ACOPs
- COP series
- HSE information sheets
- Sample forms and records
- Latest news and hot topics
- Telephone support helpline

The even better news is that you can enjoy **CRONER-i Health & Safety Expert** on a five-day FREE trial – with no obligation to buy.

Keep out of harm's way and register for your free trial now at *www.croner.co.uk/hstrial*

Notes

Notes

Notes

Notes

Notes

Notes

Notes